STARK LIBRARY JUN - - 2022
DISCARD

Bizarre Beast Battles

BOA CONSTRICTOR VS. GRIZZLY BEAR

By Charlotte Herriott

Please visit our website, www.garethstevens.com. For a free color catalog of all our high-quality books, call toll free 1-800-542-2595 or fax 1-877-542-2596.

Library of Congress Cataloging-in-Publication Data

Herriott, Charlotte, author.
 Boa constrictor vs. grizzly bear / Charlotte Herriott.
   pages cm. — (Bizarre beast battles)
 Includes bibliographical references and index.
 ISBN 978-1-4824-2780-6 (pbk.)
 ISBN 978-1-4824-2781-3 (6 pack)
 ISBN 978-1-4824-2782-0 (library binding)
 1. Boa constrictor—Juvenile literature. 2. Grizzly bear—Juvenile literature. 3. Animal behavior—Juvenile literature. 4. Animal weapons—Juvenile literature. I. Title. II. Title: Boa constrictor versus grizzly bear.
 QL666.O63H47 2016
 590—dc23

2014048080

First Edition

Published in 2016 by
Gareth Stevens Publishing
111 East 14th Street, Suite 349
New York, NY 10003

Copyright © 2016 Gareth Stevens Publishing

Designer: Katelyn E. Reynolds
Editor: Therese Shea

Photo credits: Cover, pp. 1 (boa constrictor), 12 fivespots/Shutterstock.com; cover, p. 1 (grizzly bear) Critterbiz/Shutterstock.com; cover, pp. 1–24 (background texture) Apostrophe/Shutterstock.com; pp. 4–21 (boa icon) Roman Sotola/Shutterstock.com; pp. 4–21 (bear icon) Dashikka/Shutterstock.com; p. 4 Paul Whitten/Science Source/Getty Images; p. 5 Audrey SniderBell/Shutterstock.com; p. 6 Mendel Productions/Shutterstock.com; p. 7 oksana.perkins/Shutterstock.com; p. 8 (boa constrictor) jbmake/Shutterstock.com; p. 8 (map) xZise/Wikipedia.com; pp. 9 (grizzly bear), 13 Dennis W. Donohue/Shutterstock.com; p. 9 (map) Cephas/Wikipedia.com; p. 10 Photo Researchers/Science Source/Getty Images; p. 11 Jim David/Shutterstock.com; p. 14 Mint Images - Frans Lanting/Getty Images; p. 15 Barrett Hedges/National Geographic/Getty Images; p. 16 Anton_Ivanov/Shutterstock.com; p. 17 Visuals Unlimited, Inc./Mary Ann McDonald/Getty Images; p. 18 Pete Oxford/Minden Pictures/Getty Images; p. 19 Protasov AN/Shutterstock.com; p. 21 (boa constrictor) Michael Lustbader/Science Source/Getty Images; p. 21 (grizzly bear) Scott E Read/Shutterstock.com.

All rights reserved. No part of this book may be reproduced in any form without permission in writing from the publisher, except by a reviewer.

Printed in the United States of America

CPSIA compliance information: Batch #CS15GS: For further information contact Gareth Stevens, New York, New York at 1-800-542-2595.

CONTENTS

Big, Bad Boa 4

Great Grizzly 6

Snake vs. Bear 8

Sizing Up the Enemy 10

Scary Chompers 12

How They Attack 14

Deadly Speed 16

Sneak Attack! 18

The Winner? 20

Glossary 22

For More Information 23

Index 24

Words in the glossary appear in **bold** type the first time they are used in the text.

BIG, BAD BOA

Boas are a family of snakes. There are more than 40 **species** of boas. None are venomous, or poisonous, but that doesn't mean they aren't dangerous. They kill by squeezing, or constricting, their prey until the prey can't breathe anymore. That's pretty nasty!

Boa constrictors are a large, heavy species of boa. They eat almost anything they can catch, such as birds, rats, mice, monkeys, and wild pigs. Then they open their jaws wide to swallow the prey whole.

Boa constrictors are **CARNIVORES** that hunt at night and usually live alone.

GREAT GRIZZLY

Grizzly bears are huge brown bears with a hump on their shoulders. They're named for their "grizzled" hair. That means their fur has silver or white tips.

Grizzly bears are **omnivores**. They mostly eat nuts, berries, fruits, leaves, and roots. However, they also attack and eat animals, including fish and **mammals** such as young moose. Grizzly bears sometimes eat so much they gain 3 pounds (1.4 kg) a day! Grizzlies eat carrion as well—the rotting meat of a dead animal. Yuck!

Grizzly bears usually live alone, but sometimes come together where there's a lot of food.

SNAKE VS. BEAR

What if these two tough beasts—boa constrictor and grizzly bear—were to do battle in real life? Which fierce creature would win? They live in different **habitats**, so it's not likely this would ever happen, but we can imagine!

 Mexico

 Central America

■ boa constrictor habitat

 South America

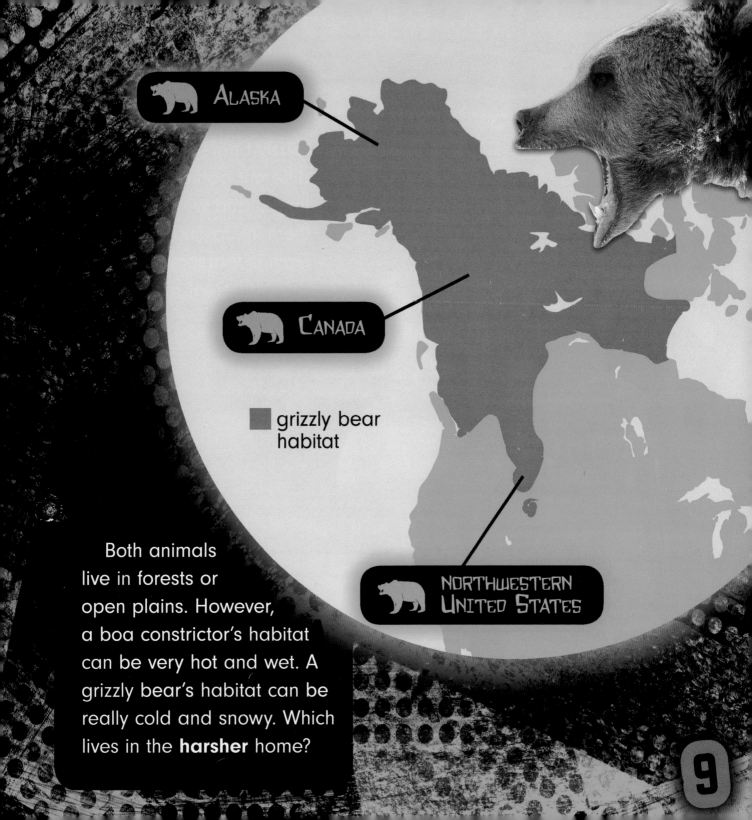

- Alaska
- Canada
- grizzly bear habitat
- Northwestern United States

Both animals live in forests or open plains. However, a boa constrictor's habitat can be very hot and wet. A grizzly bear's habitat can be really cold and snowy. Which lives in the **harsher** home?

SIZING UP THE ENEMY

Boa constrictors and grizzly bears have very different bodies, but their length and weight make them interesting **opponents**. Each is fearsome for a different reason!

up to 13 feet (4 m) long

up to 100 pounds (45 kg)

UP TO 7 FEET (2.1 m) LONG

UP TO 850 POUNDS (386 KG)

The boa constrictor is longer than a grizzly bear. However, a grizzly bear is heavier—more than eight times heavier!

SCARY CHOMPERS

Boa constrictors have sharp teeth. So do grizzly bears. They both have lots of them! A boa constrictor doesn't chew with its many supersharp teeth. It uses them to grab and hold on to prey.

 MORE THAN 100 SMALL, HOOKED TEETH

A grizzly bear has longer teeth at the front of its mouth for tearing meat. Grizzlies also have flat teeth for chewing. Which bite looks worse to you? Both mean death to prey!

42 LARGE TEETH, SOME LONG AND SOME FLAT

HOW THEY ATTACK

Boa constrictors have a special **adaptation** for finding prey. Some scales on their body sense heat, such as the body heat of animals. When the boas find prey, they wrap their thick body around it and constrict until it **suffocates**.

FIND PREY: SENSE BODY HEAT

ATTACK METHOD: SUFFOCATION BY CONSTRICTION

FIND PREY: SENSE OF SMELL

ATTACK METHOD: CHASING, KNOCKING DOWN, BITING

Grizzly bears can smell food from miles away. A grizzly bear will chase an animal, knock it down, and then bite it. Grizzlies use their claws—which can be 5 inches (12.7 cm) long!—to dig and catch animals underground. Which sounds worse: death by suffocation or by terrible bite?

DEADLY SPEED

Which animal tops the other in speed? Boa constrictors **slither** slowly, about three times slower than people walk. However, boas don't have to move fast since they **ambush** their prey. Then, they strike at incredible speed.

TOP SPEED:
1 mile (1.6 km) per hour

**TOP SPEED:
30 MILES (48 KM) PER HOUR**

Grizzlies are huge and heavy, but they're also really fast. Many of their animal prey are amazing runners, such as elk. So, grizzly bears need to be able to run at least as fast, if not faster. Grizzlies are 30 times faster than boas!

SNEAK ATTACK!

Boa constrictors' bodies can be many colors and **patterns**. Their markings are **camouflage** in their habitats. They help boas blend in and ambush prey without warning! A boa's camouflage makes up for its lack of speed.

CAMOUFLAGE PATTERNS: LINES, OVALS, DIAMONDS, OR CIRCLES

CAMOUFLAGE COLORS: TAN, GREEN, RED, OR YELLOW

CAMOUFLAGE COLOR: BROWNISH FUR

A grizzly bear isn't always easy to spot in the wild, either. The bear may wait at the edge of a forest for prey to pass by. Its brownish fur can blend in with forested habitats. Would you notice either of these animals in the wild?

THE WINNER?

Now, you've seen how boa constrictors and grizzly bears match up in several different areas. Grizzlies are heavier, but boa constrictors can be longer. Boas can suffocate their prey, but grizzly bears chew them up with their teeth. Grizzly bears move faster, but boa constrictors have excellent camouflage to sneak up on prey before they strike.

In a fight to the death, which do you think would win? Would the grizzly bite the boa in half? Or could the boa squeeze the bear to death? You decide!

BOTH ANIMALS HAVE ADAPTATIONS THAT MAKE THEM DEADLY TO OTHER ANIMALS.

GLOSSARY

adaptation: a change in a type of animal that makes it better able to live in its surroundings

ambush: to attack from a hiding place

camouflage: colors or shapes in animals that allow them to blend with their surroundings

carnivore: an animal that eats meat

habitat: the natural place where an animal or plant lives

harsh: unpleasant, difficult

mammal: a warm-blooded animal that has a backbone and hair, breathes air, and feeds milk to its young

omnivore: an animal that eats both meat and plants

opponent: someone who must be beaten in order to win a contest

pattern: the way colors or shapes happen over and over again

slither: to slide easily over the ground

species: a group of plants or animals that are all the same kind

suffocate: to die because of the inability to breathe

FOR MORE INFORMATION

BOOKS

Lanser, Amanda. *Boa Constrictor.* Minneapolis, MN: ABDO Publishing, 2013.

Owings, Lisa. *The Grizzly Bear.* Minneapolis, MN: Bellwether Media, 2012.

Somervill, Barbara A. *Grizzly Bear.* Ann Arbor, MI: Cherry Lake Publishing, 2009.

WEBSITES

Boa Constrictor
animals.nationalgeographic.com/animals/reptiles/boa-constrictor/
Read more about boa constrictors and other snakes.

Fact Sheet: Grizzly Bear
www.defenders.org/grizzly-bear/basic-facts
Find out what kinds of dangers face grizzlies in the wild.

Publisher's note to educators and parents: Our editors have carefully reviewed these websites to ensure that they are suitable for students. Many websites change frequently, however, and we cannot guarantee that a site's future contents will continue to meet our high standards of quality and educational value. Be advised that students should be closely supervised whenever they access the Internet.

INDEX

adaptation 14, 20

ambush 16, 18

bite 13, 15, 20

camouflage 18, 20

carnivores 5

chase 15

claws 15

fur 6, 19

habitats 8, 9, 18, 19

length 10, 20

omnivores 6

prey 4, 12, 13, 14, 16, 17, 18, 19, 20

sense heat 14

smell 15

speed 16, 18

squeezing 4, 20

suffocate 14, 15, 20

teeth 12, 13, 20

weight 10, 20